EXTREME MOMS

BLACKBIRCH PRESS
An imprint of Thomson Gale, a part of The Thomson Corporation

THOMSON
GALE

Detroit • New York • San Francisco • San Diego • New Haven, Conn. • Waterville, Maine • London • Munich

L. E. SMOOT MEMORIAL LIBRARY
9533 KINGS HIGHWAY
KING GEORGE, VA 22485

© 2005 Thomson Gale, a part of The Thomson Corporation.

Thomson and Star Logo are trademarks and Gale and Blackbirch Press are registered trademarks used herein under license.

For more information, contact
Blackbirch Press
27500 Drake Rd.
Farmington Hills, MI 48331-3535
Or you can visit our Internet site at http://www.gale.com

ALL RIGHTS RESERVED.
No part of this work covered by the copyright hereon may be reproduced or used in any form or by any means—graphic, electronic, or mechanical, including photocopying, recording, taping, Web distribution or information storage retrieval systems—without the written permission of the publisher.

Every effort has been made to trace the owners of copyrighted material.

Photo credits: Cover: all Corel Corporation except top left, bottom left © Digital Stock; bottom right © Royalty-Free/CORBIS; top right © PhotoDisc; interior: all pages © Discovery Communications, Inc., except for pages 1, 8, 13, 18, 34, 38 Corel Corporation; pages 4, 22, 23 © Digital Stock; page 6 © Bruce Glassman; page 12 © Royalty-Free/CORBIS; page 26 © PhotoDisc; page 30 © Terry Whittaker; Frank Lane Picture Agency/CORBIS

LIBRARY OF CONGRESS CATALOGING-IN-PUBLICATION DATA

Moms / Sherri Devaney, book editor.
   p. cm. — (Planet's most extreme)
Includes bibliographical references.
ISBN 1-4103-0390-X (hardcover : alk. paper) — ISBN 1-4103-0432-9 (pbk. : alk. paper)
1. Parental behavior in animals—Juvenile literature. I. Devaney, Sherri. II. Series.

QL762.M65 2005
591.56'3—dc22
                                                            2004019484

Printed in the United States of America
10 9 8 7 6 5 4 3 2 1

Some moms are pretty cool, but it's not all love and kisses when you're dealing with Mother Nature. We're counting down the top ten Most Extreme moms in the animal kingdom and comparing their maternal instincts to those of humans.

10 The Elephant

Kicking off the countdown is one big mother. The elephant is an extreme mom—an extremely big mom. The elephant is pregnant for a record 22 months. Then she gives birth to a 200-pound baby! She's number ten in the countdown because no other land animal gives birth to a bigger baby.

Even the average American baby is enormous if you compare the body weight of human mothers with elephant mothers. When an 8,000-pound elephant gives birth to a 200-pound baby, it's about one-fortieth of her body weight. When a 140-pound woman gives birth to a 7-pound baby, that's one-twentieth of her body weight. That means a human baby is, relatively speaking, twice the size of a newborn elephant!

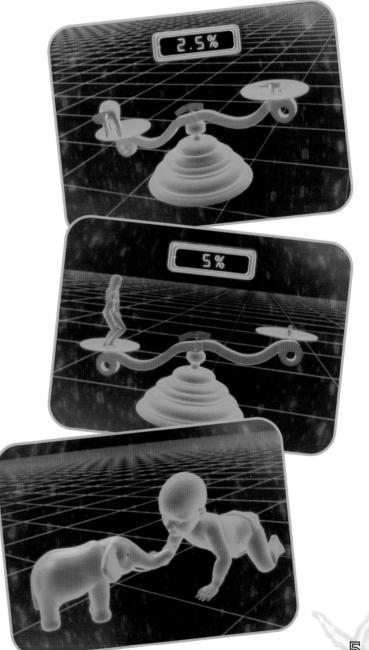

In proportion to the body weight of their mothers, human babies are actually bigger than baby elephants!

Even if proportionately elephants don't have the biggest babies in the world, these moms are still extreme. That's because in elephant families it's a case of "mother knows best." In this matriarchal society, the most extreme mom is actually a grandmother.

You won't find any males in this herd. That's because the herd is made up of grandmothers and their daughters and granddaughters.

These babies sure have it made. Their moms lead them right to the tastiest grass to munch on.

Because the males are forced out at an early age, the herd is mostly made up of this grandmother's daughters and granddaughters. With so many mothers in the herd to care for them, it's no wonder that baby elephants are constantly showered with affection.

9 The Koala

The trees of the Australian outback are home to number nine in our countdown of most extreme moms. When it comes to devotion, there's one animal that scales new heights—the koala.

She may spend up to 22 hours each day sleeping, but this is one mother who'll go to extreme lengths to look after her baby. It's not her climbing ability that makes the koala number nine in the countdown— it's her extreme digestion.

The koala is one of the pickiest eaters in the world. It only eats the highly poisonous leaves of eucalyptus trees. It manages to survive on this deadly diet thanks to a digestive system packed with special bacteria that detoxify the leaves. Koala babies are not born with the bacteria, so koala moms have to pass the bugs on. They do that by making a tasty treat of bacteria-filled droppings. A koala baby may look sweet, but it has a real potty mouth.

Boy, this koala mom sure likes to pig out! She's loading up on eucalyptus leaves so her babies will have enough of her droppings to eat.

Life would be so much easier if humans could understand what their babies are trying to tell them.

In Washington, D.C., some babies are learning to talk—not with their voices—but with their hands! Some are only nine months old, and yet they're learning to communicate with their mothers thanks to a program called Signing With Kids, run by cofounder Tegan Corrodino, who explains:

> We're teaching parents and children how to communicate with each other using American Sign Language. For example, if we teach them to sign "more" instead of waiting for them to articulate "more" it eases frustration.

"You always seem to know just what I want!" Baby Ivy is able to communicate with her mother using sign language.

Meet fifteen-month-old Ivy Mach. She already knows the signs for reindeer, ball, horse, bread, and milk, and can even tell her mom she's finished talking. It's a useful tool to reduce some of the stresses of being a mom, according to cofounder Caroline Stephan:

Unlike Baby Ivy, this koala baby has no need for sign language. When he wants something, he just jumps right on mom's back to let her know.

Every parent has been sitting at the high chair with his or her crying child and asking—do you want more? Do you want to get down? Do you want more of something I gave you ten minutes ago? Or do you want something that we haven't had yet? Just giving parents some basic signs, some food signs, or "all done" or "more" does wonders in easing the frustration between parents and child because it empowers the child to communicate.

It's a shame that koala bear babies don't know sign language. Then they'd be able to communicate just how much they love feeding on their mother's droppings.

8 The **Alligator**

Be warned! Playing with the next contender in our countdown of extreme moms can be bad for your health. Fighting her way into number eight in the countdown is the alligator. She can be deadly, especially when she is near the compost heap she calls a nest. It's easy to believe that she has no maternal instincts at all when her babies end up between those massive teeth!

Would you want to sleep on a bed of sharp teeth? That's just what baby alligators do when they rest in mama's mouth.

Doesn't every mother know not to let kids play with sharp objects? So why on Earth would this extreme mother want to eat her own babies? To figure it out, you would need to travel back 60 days.

At first glance, a pile of rotting vegetation may seem like a poor excuse for a nest. But it's this decaying compost that creates enough heat to incubate alligator eggs. This extreme mom doesn't have to touch her eggs for two months, and she makes sure that nobody else touches them either! So for researchers, checking the temperature of alligator eggs is a risky business!

This researcher better hurry if he wants to check the temperature of these alligator eggs. Mama alligator doesn't like having him around!

Although a sonogram can tell a human mother the sex of her unborn child, alligators have to wait for the eggs to hatch to know if they're boys or girls.

If the temperature in an alligator nest is less than 88°F all the eggs will be females. If it's above 91°F they'll all be males. Life would be easier for researchers if finding out the sex of the babies was as simple as getting a sonogram.

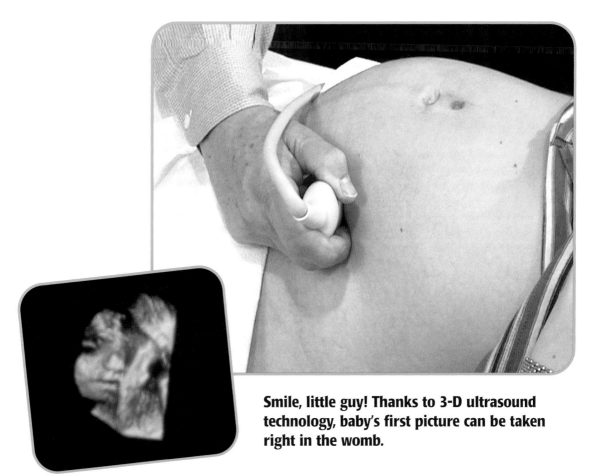

Smile, little guy! Thanks to 3-D ultrasound technology, baby's first picture can be taken right in the womb.

A medical probe emitting high frequency sound waves creates sound pictures of an unborn baby. Thanks to new sonogram technology, it's now possible to get a three-dimensional "womb with a view"! Using special 3-D ultrasound, Chicago's Dr. Jason Birnholz can paint a picture of an unborn baby—even the details of the child's face! Dr. Birnholz explains how such extreme technology may prove to be a lifesaver:

We want to know if a fetus is healthy or sick. This is the primary concern in obstetrics. So we've come to looking at the face, because it is an organ of communication. We want to be able to identify the fetus that is happy and serene in his or her environment and the fetus that is in some way troubled and in an inadequate environment. If a fetus is sick or distressed you want to get it out as soon as possible. They'll do much better in the outside world than they will inside.

Alligator mothers also use sound to find out if their babies are ready for the outside world. When they hear muffled cries from inside the compost heap, it's time to start digging.

The alligator is number eight in the countdown because she's not eating her babies, she's carrying them to safety in her armor-plated mouth. Who said you had to have warm blood to be warmhearted?

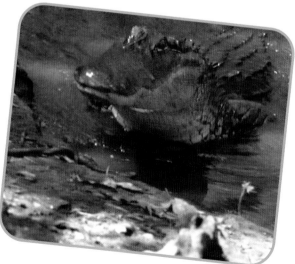

This alligator mom gives her kids the safest ride possible, right between her jaws.

7 The Polar Bear

Number seven in our countdown—the polar bear—lives in a most extreme environment, and is extreme in more ways than one.

Because they are loners, it's tricky for a polar bear to find a mate and get pregnant. In fact, one male was recorded tracking a female's paw prints for more than 60 miles! But once a female polar bear has mated with the male, she is on her own.

The polar bear now has to find enough food to almost double her weight. If she doesn't gain at least 400 pounds of fat, her body will simply reabsorb the fertilized egg. All that extra weight is essential, because she disappears for the Arctic winter and won't eat again for more than two months. Inside her snow cave, the female falls into a deep sleep. She won't wake up for anything, not even to have a baby!

This pregnant polar bear has eaten enough food for both her and her growing baby to hibernate through the winter.

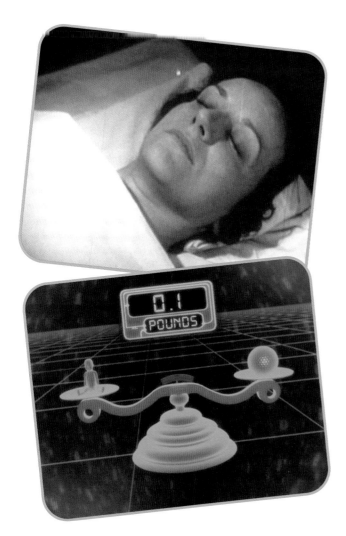

Drugs used to help women sleep through the pain of childbirth (top). If babies weighed as much as golf balls (bottom), delivery would be perfectly painless.

Doctors once tried to help women almost sleep through their labor, too. Thanks to a cocktail of painkillers, moms became so relaxed that sometimes they couldn't remember anything about the birth of their child. Although the deliveries were painless, most moms went on to suffer postnatal depression, so the "twilight sleep" treatment was abandoned.

Giving birth would be so much easier if we were like polar bears. Why bother waking up if, relatively speaking, our baby was the same size as a polar bear cub? A thousand-pound polar bear gives birth to a one-pound baby. That would be like a human giving birth to a baby smaller than a golf ball!

A polar bear cub clings to its mother's fur for warmth and protection. Until the cub is fully grown, it won't leave mama's side.

A polar bear cub is blind, toothless, and very cute. It will stay with its sleepy mother for ten weeks, feeding off her fat, which has been converted into energy-rich milk. But in this extreme landscape, the polar bear mom still has to work hard bringing up her baby. For the next two years her cub will barely leave her side.

6 The Cheetah

Our next extreme mom earned her position as number six honestly, even though she's a cheetah! The cheetah is one of the hardest working moms on the African plains because she has up to six hungry mouths to feed. That means she has to hunt every day. Unfortunately, there is no cheetah child care on the plains, and after only six weeks, her cubs start helping her hunt.

Look at mama go! By watching their mothers chase down prey, baby cheetahs learn how to hunt.

The only way the cubs can learn to hunt is by watching their mother in action, slowly stalking her victim until she's close enough to unleash her explosive speed. It's an exciting time for the cubs—in fact, sometimes too exciting for a cub to stay quietly away and not ruin mom's chances of catching her prey. Bringing up kids isn't easy, and not just for cheetahs.

Kelly Vlahov almost never gets a break. Caring for her quadruplet daughters takes up most of her time.

Meet a most extreme human mom. In one year, Kelly Vlahov and her husband will change more than 9,000 diapers, make 1,275 lunches, and give 1,040 goodnight kisses. That's because Vlahov has quadruplets.

In America today the chances of a pregnant woman having a multiple birth are higher thanks to the growing use of assisted reproductive techniques. But if having one baby is difficult, just imagine how much work is involved in caring for quadruplets! Vlahov explains:

> *The hardest thing is that your day is so constant. There is never a break, and there is always something that needs to be done, such as getting their food, drink, diapers or clothing. It's a lot to handle with one child; then multiply*

that times four. So I would say the hardest thing is trying to keep everything organized. What makes it all worthwhile is the "I love you, Mommy." And the fact that it's times four makes it even better. You're so proud of them because every day they grow and they learn new things, even if it's just coming home from nursery school singing a new song. Everything about them is amazing.

The cheetah mom is also amazing, for as soon as her cubs have learned to hunt, she leaves to start a new family of fast run-

Although Vlahov will spend years raising her quadruplets, this cheetah mom will leave her cubs just as soon as they learn how to hunt.

5 The Orangutan

On the Indonesian island of Sumatra, hidden deep in the rain forest, number five in our countdown of most extreme moms is a real swinger—the orangutan.

This ultimate high-rise mom spends her entire life far above the dangers on the forest floor. But hanging around in trees isn't easy, because this mom's the ultimate homemaker. Just like a human mother, she always has work to do around the house.

Imagine if moms were paid for all the jobs they carry out every day. Your average mom has to be a child care worker, recreation worker, chef, food service worker, property manager, housekeeper, animal caretaker, registered nurse, management analyst, computer systems analyst, financial manager, driver, psychologist, general office clerk, and social worker. All rolled into one!

It's been estimated that if moms were paid for everything they did, it would add up to an annual salary of more than $500,000.

If human moms were paid for all the work they do, they'd earn an awful lot of money!

When it comes to homemaking, though, not even the busiest mom is a match for the orangutan. She's number five in the countdown because she builds a new home every single night. That means in the course of her life, she'll build more than 30,000 homes from scratch!

A baby orangutan clings tightly to its mom as she gathers tree leaves for their nest.

At the end of the day this busy mom has no time to hang around. As high as 60 feet above the ground she begins work on her nest. By bending over branches she makes a firm base and then tucks in the smaller twigs to make a springy mattress.

The finished sleeping platform is so safe that this extreme mom can finally take the weight off her arms and spend some quality time with her baby.

Tucked in safely for the night, mom and baby spend some quiet time together in their nest.

4 The Red Knobbed Hornbill

The island of Sulawesi in Southeast Asia is home to the most extreme stay-at-home mom on the planet. Few mothers spend more time literally stuck in the house than the animal that's sealed its position at number four in the countdown.

Welcome to the world of the female red knobbed hornbill. For two months this devoted mom will be imprisoned on her nest inside the trunk of a hollow tree. But there's a good reason for her confinement. Sulawesi may look like a peaceful paradise for a nesting hornbill, but the local monitor lizards are much more than just social climbers. So to keep out hungry neighbors the hornbill mom seals herself inside a hollow tree. She's number four in the countdown because her building material is made from her own droppings! It sets like smelly concrete, leaving only a tiny window to the outside world, where her mate is roaming around and stuffing himself full of fruit. Luckily he keeps his mate well supplied with a steady supply of food leaving her to concentrate on laying her egg and raising their chick.

This female red knobbed hornbill seals the entrance to her nest inside a hollow tree using her own droppings.

Look out below! A hornbill mom sticks her rump out of her hiding place and relieves herself.

The only problem about living sealed in a tree trunk is that what goes in, must come out. The steady diet of figs does wonders for the bowels, so this extreme mom and her baby quickly learn the techniques of basic nest hygiene.

Human moms have had to find different techniques. In Elizabethan times, most babies were treated to a fresh cloth diaper once every four days. Luckily today we know a little bit more about hygiene. Unfortunately, along with our understanding of hygiene came a huge increase in laundry.

Driven to desperation by constant laundering of cloth diapers, Idaho housewife Marion Donovan developed the world's first disposable diaper. Disposable diapers have revolutionized baby care. They are so popular that the average baby goes through 3,000 diapers a year.

That means in America alone, every single day, babies fill 750,000 tons of diapers. So every day, America is getting a pile of dirty diapers as tall as a ten-story building.

The female hornbill may not have to worry about diapers, but after 60 days sealed in a tree with her baby, she's finally had enough. The chick will stay in the nest, but now this doting mom finally gets to spread her wings and fly the coop.

Disposable diapers help moms keep their babies' bottoms nice and clean (top). American babies use 750,000 disposable diapers each day (bottom).

3 The Elephant Seal

On the islands of the sub-Antarctic, number three in our countdown of extreme moms lives a soap opera life. Huge guys will fight for her. She'll live in a harem, lose immense amounts of weight, have a child, and then run away to sea. All this in one blockbuster season for the elephant seal!

**It's no wonder this elephant seal mom looks wiped out.
She spends eleven months of every year pregnant!**

The show begins when our extreme mom comes ashore to meet her leading man who is at least four times her size! She's no lightweight either. On a good day she can weigh about the same as six refrigerators, tipping the scales at 1,700 pounds!

This extreme mom is pregnant for eleven months of every year. That's eleven months of having to eat for two—that is, two pounds of blubber. She needs to put on a couple of pounds every day so that she'll get big enough to play the part of mother to her pup. Such extreme weight gain doesn't worry these big moms like it does human moms.

An elephant seal mom barks at an intruder as she protects her pup with her flipper.

Although pregnant women may feel like the size of elephant seals, on average they will gain around only 30 pounds during pregnancy. About 20 pounds of that extra weight is lost in the first month after having the baby.

The elephant seal mom is number three in the countdown because in the month after having her pup she loses a third of her body weight. That's nearly 600 pounds! All the fat she put on before the birth gets converted into milk, and lots of it. She loses 20 pounds every day that she feeds her pup! That's because elephant seal milk has 15 times more fat than human breast milk, and 25 times more than cow's milk.

At four weeks old elephant seal pups are four times heavier than they weigh at birth, and their poor moms haven't eaten a thing. After a month of their mom's dramatic devotion, the pups will be able to look after themselves.

This elephant seal mom can sleep easily as her baby feeds on her milk. She knows her milk contains everything her baby needs to grow up big and fat.

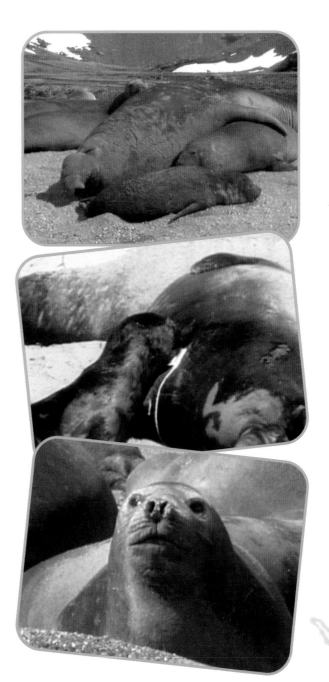

2 The Octopus

The next contender in our countdown of most extreme moms may have to look after an incredible 50,000 babies! She'll need every one of her eight arms, and in the end it will require the ultimate sacrifice. Racing into number two in the countdown is the octopus.

The octopus lives her life at a furious pace, mainly because she has to pack everything into one short year. Going from egg to adult in only twelve months means eating a lot. And she also has to find a secure place where she can lay eggs of her own, away from the menacing cod. She looks for a lair that she can easily defend, because the cod is never far behind.

Then she lays her eggs—up to 50,000 of them! She then has to regularly squirt water over them, and give them a gentle wash with her tentacles. It's a full-time job, which means she has no time to eat. It will be 40 days before her children leave home, so these extreme moms get really hungry. A starving octopus has been known to eat her own arms rather than leave her precious brood unprotected.

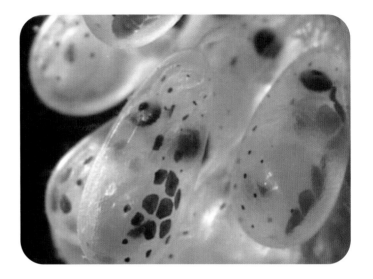

An octopus mom gives her eggs (close-up, top) some loving care with the suction discs on her tentacles (bottom).

The octopus has 50,000 good reasons to fight off cod. In the 1930s some extreme human mothers had half a million reasons to make babies.

When eccentric Canadian lawyer Charles Miller wanted to see just what extreme things people would do for money, he left a half million dollars in his will to the Toronto woman who had the most babies in the ten years following his death. A decade later, four families had tied for first place, with nine children each. This left the Canadian courts with the task of dividing the money or invalidating the will as degrading under Canadian standards of public morals.

A hungry cod looks for octopus eggs (top). Canadian lawyer Charles Miller (bottom) once held a contest to see who could give birth to the most babies.

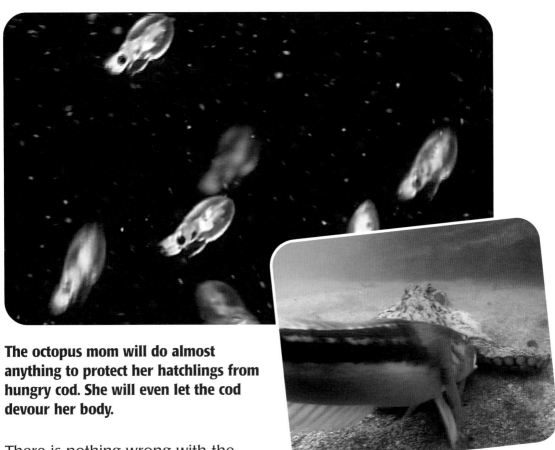

The octopus mom will do almost anything to protect her hatchlings from hungry cod. She will even let the cod devour her body.

There is nothing wrong with the morals of the octopus mom. Her babies are ready to hatch, thanks to her extraordinary devotion. But this is one mother who will never suffer from the empty nest syndrome. She's number two in the countdown because there is a price to pay for all her hard work. Her babies have finally hatched, but her old enemies are back. This time, however, she is completely exhausted and no longer has the strength to fight the cod off.

1 The Sea Louse

Some people love their moms—to death. A director named Alfred Hitchcock made a movie based on the true story of a man who loved his mommy to bits. There are some real psychos in the natural world, and the number one animal in the countdown gives birth to the worst of them all. The most extreme mom in the world is the sea louse.

A male sea louse enters his den (inset), where he's trapped a large group of pregnant females (above).

From the moment a male sea louse lures her into his lair, things go terribly wrong. Once he has her in his clutches there's no escape. She's sealed inside his burrow, with up to 25 other females who are now pregnant. Unlike humans, these extreme moms have no reason to look forward to Mother's Day.

Moms are loved so much that we celebrate them every year on the second Sunday in May on Mother's Day. It all started in 1907 when Anna Jarvis from Philadelphia successfully campaigned to set aside the anniversary of her mother's death as a special day for motherhood. She hoped that the celebration would help end all hatred between people. Nice idea, but only sixteen years later, Anna Jarvis was so upset by the commercialization of Mother's Day that she campaigned to ban the day forever! The mother sea louse would definitely agree with those sentiments.

Every year on Mother's Day children across the United States thank their moms for all their hard work with greeting cards and bouquets of flowers.

MOTHER
A word that means the world to me.

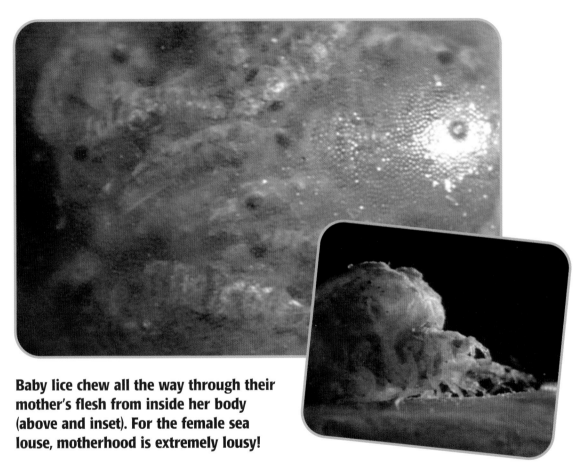

Baby lice chew all the way through their mother's flesh from inside her body (above and inset). For the female sea louse, motherhood is extremely lousy!

The mother sea louse is number one in the countdown because dozens of her psycho children have been eating her from the inside out. She dies as the withered husk of her body splits apart, unleashing her little brood into an unsuspecting ocean. This extraordinary mother has made the ultimate sacrifice for her children—so there can be no doubt, that when it comes to motherhood, the female sea louse really is The Most Extreme.

For More Information

Matto H. Barfuss, *My Cheetah Family.* Minneapolis, MN: Carolrhoda/Lerner, 1999.

John E. Becker, *The American Alligator.* San Diego: KidHaven Press, 2003.

Stephen Brend, *Orangutan.* Austin, TX: Raintree Steck-Vaughn, 2000.

Anna Clairborne, *Octopuses.* Chicago: Raintree, 2004.

Mary Ann Fraser, *How Animal Babies Stay Safe.* New York: HarperTrophy, 2001.

Eleanor J. Hall, *Polar Bears.* San Diego: KidHaven Press, 2002.

Stuart P. Levine, *The Elephant.* San Diego: Lucent Books, 1998.

Dia L. Michels, *If My Mom Were a Platypus: Animal Babies and Their Mothers.* Washington, DC: Platypus Media, 2001.

Seymour Simon, *Crocodiles & Alligators.* New York: HarperCollins, 1999.

Glossary

articulate: to speak clearly

bacteria: tiny living cells that can cause disease or do useful things, like making soil richer

blubber: the layer of fat under the skin of whales, seals, and other sea animals

compost: a mixture of decaying leaves, vegetables, or other organic matter

detoxify: the process of removing harmful chemicals from something

digestion: the process of breaking down food into a form that can be absorbed and used by the body

fetus: an unborn child in the later stages of development

incubate: to keep eggs warm for hatching

lair: the living place of a wild animal

matriarchal: governed or controlled by females

postnatal: occuring after birth

pregnant: having one or more unborn young developing within the body

prey: an animal that is hunted by another animal for food

quadruplets: four children or animals born to the same mother at the same time

sonogram: an image of a baby that is still inside the womb, which is produced by sound waves

Index

alligator, 12–17
assisted reproductive techniques, 24

bacteria, 9
blubber, 35

cheetah, 22–25
cod, 39, 40, 41
communication, 10–11
compost, 12, 14, 17

diapers, 32–33
digestion, 9
droppings, 9, 31, 32

eating, 19, 39
eggs, 14–15, 19, 31, 39
elephant, 4–7
elephant seal, 34–37
eucalyptus trees, 9

face, 16, 17
fat, 19, 21, 36–37

grandmother, 6–7

hatchlings, 41
homemaking, 27–29
hornbill, 30–33
hunting, 22, 23, 25
hygiene, 32

koala, 8–11

labor, 20
lizards, 31

mating, 19
matriarchal society, 6
milk, 21, 37
monitor lizards, 31
Mother's Day, 43–44
mouth, 17
multiple births, 24

nest, 12, 14–15, 29, 31

octopus, 38–41
orangutan, 26–29

polar bear, 18–21
pregnancy, 4, 19, 35, 36

quadruplets, 24–25

red knobbed hornbill, 30–33

sea louse, 42–45
seal, 34–37
sign language, 10–11
sleeping, 9, 19, 20
sonogram, 15–17
sound, 16, 17

talking, 10
teeth, 12, 13

weight, 5, 19, 35, 36–37

J Extreme moms.
591.563
Ext

L. E. SMOOT MEMORIAL LIBRARY
9533 KINGS HIGHWAY
KING GEORGE, VA 22485